INVASIVE BAMBOOS

*Their impact and management
in Great Britain and Ireland*

Brian Taylor Jim Glaister Max Wade

PACKARD PUBLISHING LIMITED
CHICHESTER

INVASIVE BAMBOOS
Their impact and management
in Great Britain and Ireland

© Brian Peter Taylor, James Lloyd Glaister, Paul Maxwell Wade

First published in 2021 by Packard Publishing Limited,
14 Guilden Road, Chichester, West Sussex, PO19 7LA, UK.

ISBN 978 1 85341 143 4 paperback

A CIP record can be obtained from the British Library

All photographs are by the Authors, except those credited in the Acknowledgements and List of illustrations. The dramatic image on the back cover of invading running-bamboo rhizomes under paving is by Paul Copper.

Edited by Jim Glaister
Commissioned and prepared for press by Michael Packard
Design and layout by Hilite Design, Marchwood, Southampton, Hampshire
Printed and bound in the UK by KnowledgePoint Ltd, Winnersh, Berkshire

Contents

Acknowledgements

The Authors would like to thank their families, friends and colleagues who supported and encouraged them in the creation of this book. The majority of the photographs and illustrations originated in the Authors' private and professional archives. In addition, they would like to thank the following people for taking images and allowing them to be included in this book: Alison Taylor, Paul Copper, Paul Hamilton, Kevin Gilderson, Charlie Hellard, Steph Hughes, Chris Sullivan and Winnie Wade; also Sridhar Gutam, courtesy of Wikimedia Creative Commons.

We would also like to thank George Beckett for his assistance in enabling some species to be photographed, and Bill Taylor for proof-reading this book and for his constructive comments.

1   Introduction

Bamboo is of increasing concern in Great Britain and Ireland, with some species having a detrimental impact on gardens (Fig 1.1), the built environment and biodiversity in the wild. This book aims to:

- Provide an overview of bamboo history, biology and ecology;
- Highlight the issues bamboo can cause;
- Provide useful advice for those who have planted, or wish to plant, bamboo;
- Supply practical guidance for those looking to tackle a problem with invasive bamboo.

This book offers some advice on how to go about identifying bamboos, but does not seek to provide a detailed identification guide to the various species.

Figure 1.1 Bamboo growing in a residential setting.

2 Overview

Bamboo consists of a number of hollow-stemmed species of the grass family Poaceae (sub-family Bambusoideae). The word bamboo comes from the name 'bambu', originating from the Karnataka region of south-west India. There are over a 1000 different species of bamboo with at least 30 of them commonly grown in gardens in Great Britain and Ireland. Over 20 species have invaded non-cultivated land.

Figure 2.1 *Bamboo leaves with short leaf stalks and veins running parallel along the length of the leaf.*

Figure 2.2 *A large, mature stand of bamboo.*

The bamboo plant consists of rhizomes (underground stems), uniform hollow stems (known as culms) which can range from up to two to six metres in height, flat leaves with parallel veins running the length of the leaf (mostly with a short leaf stalk) (Fig. 2.1) and, on rare occasions, flowers. A single plant may form a large colony (otherwise known as a stand) (Fig. 2.2) of many culms all linked through the rhizome network. Please refer to the Glossary for explanations of any technical terms used in this book.

A distinction is often drawn between differing types of bamboo by the way in which they grow. Generally, plants are divided into two main categories, known colloquially as 'running' and 'clumping' bamboos.

Running bamboos (Fig. 2.3) are typically invasive bamboos, spreading rapidly by growing long horizontal rhizomes that can grow vigorously into adjoining areas. These are the main source of problems in gardens or in the wider environment, as they can cause significant damage to hard surfaces

Figure 2.3 *A running bamboo with emergent shoots to the right of the main culms.*

(such as patios or terraces, drives, roads, pathways) as well as diminishing the usability and appearance of gardens and other amenity areas. They also readily cross boundaries, potentially causing the same problems in neighbouring properties.

Clumping bamboos (Fig 2.4) form a tight rootball and are less aggressive in their spread. However, they are not entirely without problems. A clumping bamboo 4.5 m tall can typically form a cluster of culms of a similar distance across, which may well cause issues with neighbours when planted too close to a boundary. Bamboo rhizome structures and the division between types are discussed in greater detail in the following chapter.

Many bamboos are able to thrive in the British climate and have few, if any, native predators or diseases that might offer natural control. Given the vigorous growth of some species and their propensity to spread, some bamboos are causing a great deal of concern, particularly in the built environment. These problems are likely to increase in the future as a result

Figure 2.4 *A clumping bamboo with two emergent shoots in the foreground.*

of climate change. The frequency of bamboos in the wider environment is increasing and a sharp and rapid expansion, known as an 'exponential growth phase', has been recorded for some species. In a number of countries with a similar climate to Great Britain and Ireland (e.g. north-western USA) invasive bamboos, such as the popular golden bamboo (*Phyllostachys aurea*), have already caused considerable environmental degradation. Control measures have proved both costly and time-consuming. Invasive bamboos have also been recorded causing environmental issues across other parts of the USA and in Australia.

Today in Great Britain and Ireland, bamboos are a common cause of gardening problems and disputes between neighbours. The environmental threat is also real. The spread of bamboos into the wild is expected to cause further difficulties in the future, particularly in the displacement of native species and by restricting access.

3 Running and clumping bamboos

Bamboos generally have one of two different types of rhizome and are usually distinguished using the colloquial terms 'running' and 'clumping' bamboo. The division between the types is somewhat arbitrary, as some types may be clump-forming in poor soil conditions, but become running in more fertile soil (*Phyllostachys* species are known for doing this, for example). There is also a third form of rhizome system called 'amphimorph', which demonstrates both clumping and running characteristics at the same time. However, these are very rare and only three types of bamboo are known to grow in this manner: most notably members of the *Chusquea* genus from South America.

The division between 'clumper' and 'runner' is therefore for guidance only. Generally speaking, it is the running bamboos that are the most invasive, but clumping bamboos can also cause problems if inappropriately planted, positioned or are poorly maintained (or if soil conditions result in them developing more of a running habit). Although running bamboos are often referred to as 'invasive' while the clumping bamboos are regarded as 'non-invasive', most bamboos have the potential to invade. It is just that some species are much more invasive than others.

The most commonly found bamboos in Great Britain and Ireland are listed in Table 3a.

Running Bamboos	Clumping Bamboos
Arundinaria, Bashania, Chimonobambusa, Clavinodum, Hibanobambusa, Indocalamus, Phyllostachys, Pleioblastus, Pseudosasa, Sasa, Sasaella, Sasamorpha, Semiarundinaria, Sinobambusa, Yushania*	*Bambusa, Chusquea*, Dendrocalamus, Drepanostachyum, Fargesia, Himalayacalamus, Shibataea, Thamnocalamus*
* Members of these genera may be clumping or running (or both) depending on soil type, microclimate and species.	

Table 3a: Running and clumping bamboo species commonly found in Great Britain and Ireland.

Running bamboo

Running bamboos have what is called a 'leptomorph' rhizome system (Fig. 3.1). The rhizomes push horizontally through the soil maintaining a depth or level. They are typically thin in appearance, jointed and can reach considerable lengths. Some species are capable of producing rhizome growth of up to six metres in a growing season, though distances approaching one to one and a half metres are more common. Culms (and sometimes new rhizomes) form from lateral or side buds along the rhizome length and can emerge from anywhere along the rhizome. Running bamboos are very efficient at colonising adjoining areas (Fig. 3.3). These characteristics make such bamboos very difficult to control.

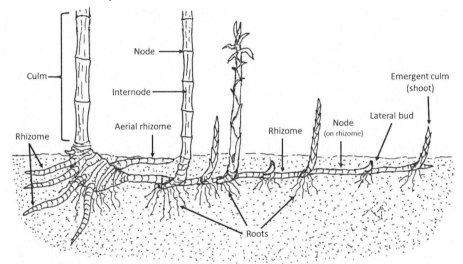

Figure 3.1 *Leptomorph rhizome system.*

Running bamboos consist of many different genera and species and these can range from a third of a metre to nearly 25 m in height. The rate of spread, as well as the density of the growth, varies from species to species.

Clumping bamboo

Clumping bamboos have what is called a 'pachymorph' rhizome system (Fig. 3.2). The rhizomes grow only short distances each year. Buds are formed on the tip of the underground rhizome, which turns upward and forms a new culm immediately (as opposed to running bamboos that can grow up to a metre or more before producing culms). This growth habit allows new shoots to emerge in close proximity to the plant. The rhizomes are typically quite thick. In mature plants, the concentration of culms can be very dense. Clumping bamboos spread slowly and are relatively easy to control.

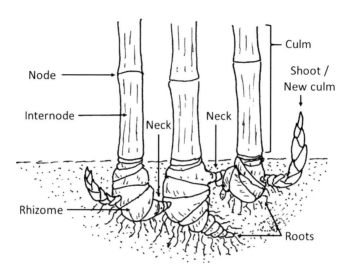

Figure 3.2 Pachymorph rhizome system.

Clumping bamboos consist of many different species and can typically range from three to four and a half metres in height. If left uncontrolled, some clumping species can spread up to a metre or more beyond the centre of the plant.

Figure 3.3 Encroachment of running bamboo from beyond a neighbouring boundary fence.

4   Recognising bamboos

The purpose of this book is to consider the problems caused by invasive bamboos, and provide methods for their control and management, rather than the identification of individual species. The control measures listed later in this book should be suitable for most, if not all, species. However, some general means of identifying bamboo can be discussed.

On the whole, bamboos are quite characteristic and readily distinguished from other grasses and plants (Table 4a). That said, there are several other plants that are often misidentified as bamboos (Tables 4b-4g). The term 'bamboo' includes a number of genera and a larger number of species and there is a great deal of variation between species (see Appendix II: Tips for identifying the bamboo on site). In addition to this, there are considerable numbers of cultivars and natural varieties which may alter the plant's appearance to the untrained eye, as well as new imports of bamboo species occurring in both the licensed and unlicensed movement of plants. As a general rule, plant identification is typically confirmed at the flowering stage, but this is an unreliable method for bamboo identification. Intervals between flowering vary greatly from species to species and can range from decades to as long as 140 years (see Chapter 6).

Variations between bamboo species to note include:

- **Height:** can be variable, dependent on species, maturity and growing conditions. Bamboos can be small; for example, dwarf bamboo (*Pleioblastus pygmaeus*) rarely reaches one metre tall.

- **Leaf shape**: not all bamboos have the typical bamboo leaf shape. Some have quite broad leaves. For example, broad-leaved bamboo (*Sasa palmata*) has leaves of three to nine centimetres in width.

- **Leaf colour**: leaves can range from the typical bright green of most bamboos through to dark greens. Variegated forms also exist.

- **Culm or stem colour**: bamboo culms range from greens and yellows to browns and blacks.

A useful feature that all bamboos share, and which can help distinguish them from many other plants (see Figs. 4.1-11), is the hollow culm between the nodes. On occasion, when the bamboo is immature or has a stem diameter of less than one centimetre, it may be difficult to determine whether the stem is hollow. The green-yellow colouration of some willows and their long narrow leaves could lead to confusion with bamboos.

Some of the knotweed family of plants can look like bamboo with their hollow, segmented stems – Japanese knotweed (*Reynoutria japonica*), for instance, is called Mexican bamboo in some parts of the United States. However, there are essential differences in the leaves. Hold up a leaf to the light and in all bamboos the veins can be seen to run lengthways and are parallel to each other. Those of the knotweeds are branching, spreading out at angles across the leaf. In addition, the leaves of various knotweed species are generally shield-shaped or heart-shaped as opposed to the long pointed leaves of bamboos (Himalayan knotweed being the exception).

Specific plants with hollow stems that are sometimes confused with bamboos, and pose more of a challenge to distinguish, are summarised in pages 16 to 23.

	Growth form	Stems and leaves	Flowers	Spread	Habitat
Bamboos (various species)	Tall, perennial, evergreen grasses (up to 6m), typically forming large colonies if left unmaintained.	Hard hollow canes with nodes. Remain green through winter. Characteristic stem sheaths. Leaves typically green, flat, mostly with short leaf stalk and tapering to pointed tip.	Rarely produce flowers.	Rhizomes, often extensive. Can seed if flowers are produced.	Grown in gardens and increasingly seen outside cultivation in a range of habitats, including roadsides, tracks, by lakes/streams and in damp shaded environments.

Table 4a Features of bamboo species.

While many plants are difficult to recognise during the winter when they lose their leaves, most bamboos are evergreen and retain their leaves.

Figure 4.1 *Aerial rhizomes emerging from the soil for a short section.*

Figure 4.2 *Auricles at the top of the sheath (indicated by arrows).*

Figure 4.3 *Blades growing from new bamboo shoot.*

Figure 4.4 *The nodes (segment joints) are typically swollen on bamboo. Leaf branches grow from the nodes and are themselves noded.*

Figure 4.5 *Bamboo culms showing nodes (segment joints) and internodes (sections between nodes).*

Figure 4.6 Running bamboo rhizome.

Figure 4.7 Clumping bamboo rhizome.

Figure 4.8 Section of running rhizome showing fibrous roots.

Figure 4.9 Sheath on Phyllostachys aureosulcata 'Spectabilis' prior to drop.

Figure 4.10 Sheath scar (immediately below the node) after the sheath has fallen off.

Figure 4.11 *Fallen sheaths.*

Figure 4.12 *Sulcus (vertical groove on internode) indicated by the red arrow. In some species, the sulcus can appear on alternating internodes, as indicated by the yellow arrows.*

	Growth form	Stems and leaves	Flowers	Spread	Habitat
Common reed (*Phragmites australis*) (Fig. 4.13)	Tall, perennial grass up to 3.5m in height, typically forming extensive beds.	Hard, hollow, cane-like dead stems with nodes persisting through the winter. Leaves greyish-green, flat, up to 3cm wide and 20-60cm long.	Plumes, 15-40 cm, deep purplish-brown, becoming lighter as plumes form, drooping gracefully to one side. Appear late July to mid-October.	Spread mainly by extensive rhizomes, but does set seed.	A wide range of wetland habitats including ditches, marshes, swamps, fens by rivers, lakes and ponds. Rarely planted in gardens, though it is available from some garden centres, sold as a pond plant.

Table 4b *Features of common reed.*

Figure 4.13 *Common reed.*

	Growth form	Stems and leaves	Flowers	Spread	Habitat
Pampas grass (*Cortaderia selloana*) (Fig. 4.14)	Tussock-forming perennial grass (up to 3 m).	Hollow, up to 3 cm across. Dead stems persist through the winter. Leaves grey-bluish green to green, 1-3 m long and 3-8 cm wide, with serrated edge and a V-shaped cross-section. Distinct mid-vein running lengthwise.	Very large 30-100 cm long, showy with silvery-white plumes. Appear August to November.	Seeds, dispersed by the wind.	Grown in gardens and used in landscape planting. Also found on roadsides, railway em-bankments, and particu-larly coastal sites, such as cliffs, dunes and waste ground.

Table 4c Features of pampas grass.

Figure 4.14 Pampas grass.

	Growth form	Stems and leaves	Flowers	Spread	Habitat
Giant reed (*Arundo donax*) (Fig. 4.15)	Very tall (up to 6m) perennial clump-forming grass.	Sturdy, hollow, cane-like stems, 2-3cm in diameter. Dead stems persist through the winter. Leaves pale, blue-green, with heart-shaped, hairy, tufted base, 2-6cm wide at the base, tapering to a fine tip. Up to 60cm in length.	Large upright, feathery plumes, light green-whitish to silvery purple, 40-60cm long. Appear September.	Extensive network of rhizomes, bearing fibrous tap roots. Seeds are rarely fertile.	Grown in gardens and occasionally found as a casual.
Chinese silver-grass (*Miscanthus sinensis*) and **Amur silver-grass** (*M. sacchariflorus*) (Figs. 4.16 and 4.17)	Tall, deciduous or evergreen, reed- or cane-like perennial grasses forming clumps or tussocks. Usually 1-2m tall.	Solid, sometimes hollow, 3-10mm in diameter. Leaves broad, 8-12mm wide, 30-60cm long. Sometimes narrow, flat, and rolled in the bud.	Open, fan-shaped branches 20-30cm long, appear August to November, brownish-red, ageing to pink and silver. Persist through winter.	Chinese silver-grass: seeds dispersed by wind. Amur silver-grass: rhizomes.	Grown in gardens and occasionally found as a casual.
Himalayan honeysuckle (*Leycesteria formosa*) (Fig 4.18)	Deciduous shrub with drooping, branched stems. Usually 1-2m tall.	Up to 2m tall, hollow with nodes, remaining green over winter. Only last for 2–5 years. Leaves opposite, dark green, 6-18cm long and 4-9cm wide, with wavy margin and long point at the tip.	Creamy-white to purple in hanging tassels. Appear July to August.	Seeds produced in purple to purple-red berries. Spread by birds.	Grown in gardens and also found in woods, hedgerows, shrubberies and on waste ground.

Table 4d Features of giant reed, silver-grass and Himalayan honeysuckle.

Figure 4.15 Giant reed.

Figure 4.16 Chinese silver-grass.

Figure 4.17 Amur silver-grass.

Figure 4.18 Himalayan honeysuckle.

	Growth form	Stems and leaves	Flowers	Spread	Habitat
Japanese knotweed (*Reynoutria japonica*) (Fig. 4.19)	Tall, herbaceous perennial (up to 3 m), typically forming large colonies (stands) if left uncontrolled.	Hollow, noded, green with pink/purple speckles, can reach diameters of up to 40 cm, hard, cane-like, dead stems persist through the winter. Leaves shield-shaped, up to 12-14 cm long, alternate, creating zig-zag pattern to stem.	Small, creamy-white, hanging in clusters. Appear in late summer/early autumn	Extensive network of rhizomes. Rhizomes and crowns are regenerative and thrive on disturbance. Seeds are rare and do not survive the winter.	Thriving mainly in a wide range of urban habitats, particularly on transport routes (by roads, waterways, railways), also found in gardens – usually from regenerated fragments, rhizome spread or historic planting.
Giant knotweed (*Reynoutria sachalinensis*) (Fig. 4.20)	Tall, herbaceous perennial (up to 4-5 m), typically forming large colonies (stands) if left uncontrolled.	Hollow, noded, stems mature to a greenish-brown colour, leaf stems pinkish-red or green. Dead stems persist through the winter. Leaves are an elongated heart shape, up to 40 cm long and 27 cm wide, with pointed tip, crinkled appearance and fine white hairs on the undersides. They grow alternate, creating zig-zag pattern to stem.	Green-white, hanging off the plant in dense clusters (panicles). Appear in late summer or early autumn.	Extensive network of rhizomes. Can spread by seed, rhizomes or vegetative regeneration arising from human disturbance.	Not as widespread as Japanese knotweed, but can still be found in almost any habitat.

Table 4e Features of Japanese knotweed and giant knotweed.

Figure 4.19 *Japanese knotweed.*

Figure 4.20 *Giant knotweed.*

	Growth form	Stems and leaves	Flowers	Spread	Habitat
Hybrid knotweed (*Reynoutria x bohemica*) (Fig. 4.21)	Herbaceous perennial, typically formed by interaction of giant knotweed and Japanese knotweed (though other hybridisations can occur) – height can vary from 2.5 m to 4 m. Forms large colonies (stands) if left uncontrolled.	Hollow, sturdy, noded, green with pink/purple speckles, hard cane-like, dead stems persist through the winter. Leaves alternate, creating zig-zag pattern to stem. Form varies from shield-shaped to heart-shaped, size variable (up to 25 cm long and 18 cm wide), typically darker green than Japanese knotweed and crinkled.	Creamy-white, densely-packed, appear in late summer or early autumn, often found in a male form, with flower clusters upright rather than hanging down off the stem.	Extensive network of rhizomes. Can spread by seed, rhizomes or vegetative regeneration arising from human disturbance.	Not as widespread as Japanese knotweed, but increasing in number and can still be found in almost any habitat.

Table 4f Features of hybrid knotweed.

Figure 4.21 Hybrid knotweed, in this case formed by the interaction of giant knotweed and Japanese knotweed.

	Growth form	Stems and leaves	Flowers	Spread	Habitat
Himalayan knotweed (*Persicaria wallichii/ Koenigia polystachia*) (Fig. 4.22)	Tall, herbaceous perennial (1.8-2 m).	Reddish-green, noded, solid, fine white hairs just below nodes, typically pink-red where leaves branch from main stem, brown sheaths persist at nodes. Leaves elongated and slender, tapering to pointed tip, but variable in shape and length (from 10-20 cm), usually slightly crinkled along the edges, short hairs on the undersides, veins and leaf edges.	White, sometimes pink, with yellow centres, small, five-petalled, growing in upright, loose clusters around 20-35 cm long. Appear between mid-summer and late autumn.	Spreads by rhizomes, seed, or vegetative regeneration arising from human disturbance or flood events. Present in Great Britain and Ireland in both male and female forms, as the flowers are actually hermaphrodite (i.e. contain both male and female parts)	Can be found in a wide range of habitats, but prefers moist soils on banks of rivers, streams, canals and other waterways, as well as areas of the floodplain and shallower sections of stream beds.

Table 4g Features of Himalayan knotweed.

Figure 4.22 Himalayan knotweed.

5 Origins, history and uses of bamboos

Bamboos are found as native plants on most continents, with the exception of Europe and Antarctica.

Bamboos are and have been used for many purposes over the years, including scaffolding, clothing, built structures, roads, food, animal fodder, medicinal purposes, furniture, garden supports, beer, jewellery, toys and tools. As such, it is a group of plants that plays an important part economically and has a significant role in many countries.

In Europe, all bamboos are introduced exotic species and are principally planted as garden plants or for their aesthetic qualities in soft landscaping. There have been several waves of bamboo introductions since the late eighteenth century.

Bamboos introduced to Great Britain and Ireland from this earliest period had their origin in south-east Asia and were mainly only suitable for hot houses or very sheltered areas. However, the late nineteenth century saw the

Figure 5.1 *Bamboo canes repurposed as a barrier screen.*

Figure 5.2 *Bamboo planted as an ornamental feature in a garden.*

introduction of some hardier types of bamboo, with further introductions of extremely hardy varieties occurring from the late twentieth century onwards. The Japan-British Exhibition of 1910 was an example of early interest in bamboo in Britain. This exhibition included two major Japanese gardens, both featuring bamboo, and attracted over eight million visitors in just a six-month period. More recently, the 1990s saw bamboo being popularised in books, magazines, the press and on television as a number of garden experts actively promoted planting them. Bamboos remain a popular choice of plant for the garden today.

There is a great deal of advice available on how to contain newly-planted bamboos, with many nurseries promoting the use of root-barrier membranes or containing bamboos in pots. In reality, the physical effort of excavating a suitable trench for a root barrier means they are rarely installed at sufficient depths or distances along a boundary to provide adequate or suitable control once the bamboo is established. More often, membranes are not installed at all. As with most plants, bamboos require maintenance. Such tasks can include thinning out the culms, limiting the height and severing rhizomes to prevent spread. However, such tasks are seldom performed adequately, if at all.

See Chapter 11 for a more detailed discussion of bamboo containment and maintenance.

6 Biology, ecology and dispersal

Most bamboos are sold by the plant-nursery trade as containerised plants, which may have several shoots and some rhizome material. Nurseries are often unclear about whether the plants sold are running or clumping types and may not stress the importance of maintenance. This lack of advice may lead to poor plant selection, an inappropriate choice of planting location and subsequent problems.

The typical life cycle of bamboos involves the development of a large mature stand or colony, which can extend hundreds or thousands of square metres for some species (Fig. 6.1). Flowering, seed formation and then typically death (or severe weakening) of the parent plant may follow.

Dispersal of bamboos is largely dependent on people either accidentally or deliberately spreading sections of rhizome, for example in garden waste and topsoil. Propagation is more likely where a rhizome is relatively young (i.e. less than two years old), has at least two dormant buds and is dispersed in the early spring. Distribution by seed is not a significant cause of spread at the time of writing. When existing areas of bamboo mature, seed formation and dispersal may well occur – although at what level is currently uncertain.

Once new plants are established, running bamboos will create extensive rhizome networks around them. These horizontal networks will develop vertical aerial growth (i.e. the culms) and the visible bamboo area will increase in stature and range. Culms grow during the spring and early summer and reach their full height within two to three months of emerging from the rhizome. Individual culms will not grow any taller than in their first season. Instead, the bamboo will gain height by producing new culms each year that will grow taller than those of the previous season. This process will continue until the bamboo achieves its maximum height. Culm height and rhizome spread are dependent on several factors, including species, soil type, climate, availability of water and age. Culms become woodier as they age and lose their colour as they get older (Fig. 6.2). On average, the typical life span of a culm is around ten years. Established plants grow at a faster rate than more newly planted ones.

Figure 6.1 *A well-established stand of bamboo.*

During late summer and autumn, the rhizome biomass expands in preparation for the production of the following year's culms.

Bamboo rhizomes are typically strong, woody, not easily broken and with pointed tips that enable them to penetrate a wide range of surfaces and materials. Rhizomes are naturally shallow-growing, and in running species they are usually found no more than 30 cm beneath the surface of the ground. In some species, aerial rhizomes are often visible on or above the surface. Though shallow, the rhizome mass can be extremely dense. In some cases, if bamboos find insufficient nutrients or water in shallow soils, rhizomes can grow to a greater depth, though rarely more than a metre. Unlike the tough, woody rhizomes, bamboo roots are thin and fibrous and can be typically found to a depth of up to one metre.

Bamboo species exhibit several types of flowering (Fig. 6.3), which include gregarious, sporadic and annual flowering.

Figure 6.2 Culms of different ages. Those with least colour are the oldest; those with sheaths still attached are the youngest.

Gregarious flowering is when an entire population of a bamboo species flowers at the same time. This happens at widely spaced intervals, typically between 60 to 130 years. The flowering cycle is predetermined by the plant's genes, triggering plants from the same population to flower at the same time, even when they have been transported and grown for some years in another part of the country or even on another continent. The plants will flower, set seeds and typically die or be severely weakened.

Sporadic flowering appears to be determined by environmental factors and does not necessarily occur at the same time as in other members of the same bamboo population.

Annual flowering bamboos, which flower every year, are relatively rare but do exist. Bamboo plants with sporadic or annual flowering do not normally die after coming into bloom.

Encroachment by bamboo into an adjoining area may occur directly through natural rhizome spread, extending an existing colony or stand, or from fly-tipped material. As with other rhizomatous species, bamboos can regenerate from fragments of rhizome – see 'Escape into the wild' below. There is an additional dispersal mechanism which could well cause problems in the future: soil recycling, in which soils are removed from a site, screened and sold on. While the screening process may well remove much of the rhizome material, it could leave viable propagules for dispersal onto new sites. Not only is there the risk of spread of regenerative rhizome fragments, if bamboo seeds are present in the soil, they will be unlikely to be screened out. Unlike other problematic plants, such as Japanese knotweed, soils infested with bamboo propagules are not classed as controlled waste and their disposal is not (at the time of writing) covered by any legislation or regulation.

Figure 6.3 Bamboo in flower.

Escape into the wild

Although bamboos are not native to Great Britain or Ireland, a number of bamboo species have 'escaped' from gardens and have established in the 'wild'. Here they are often left to spread unmanaged, having a detrimental impact on native biodiversity.

Figure 6.4 *Bamboo established in the wild on Exmoor.*

Bamboo encroachment of non-cultivated areas has become a problem for Europe and the United States. Golden bamboo (*Phyllostachys aurea*), a bamboo widely available and planted in Great Britain and Ireland, has been identified as a particular issue in many of the southern states of America as a result of its invasion and displacement of native plant species. In 2006, the Georgia Exotic Pest Plant Council deemed it to be a 'moderate' threat. South Carolina listed it as a 'severe threat' in 2008 and the Florida Exotic Pest Plant Council ranked it as a Category II invasive plant in 2009. The spread of this bamboo has resulted in the loss of native vegetation in Texas and Hawaii. The Texas Invasive Plant Pest Council reported the bamboo growth rates as 'aggressive', while in Hawaii researchers discovered a golden bamboo stand occupying more than an acre of a steep hillside in Kailua, Oahu, excluding most other native vegetation.

This precedent of bamboos escaping into the wild clearly demonstrates how they can affect their surrounding environment. This is a problem that needs to be officially acknowledged in Great Britain and Ireland before the same situation develops as with, for example, Japanese knotweed: that is, trying to contain the problem long after it is already too well established.

Arrow bamboo *(Pseudosasa japonica)*   Broad-leaved bamboo *(Sasa palmata)*

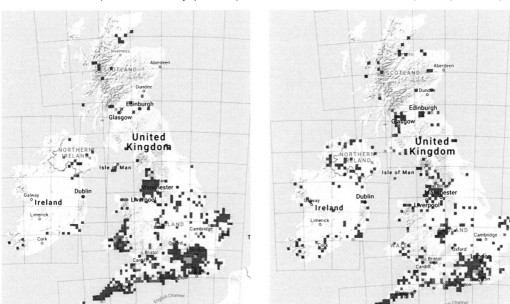

Figure 6.5. Distribution maps of two bamboo species in the wild in Great Britain and Ireland. Based on records mainly collected by the Botanical Society of Britain and Ireland (BSBI). Used by permission of BSBI, with grateful thanks.

By 1996, twenty-five species of bamboo had been recorded in the wild in Great Britain and Ireland. Table 6a summarises those species encountered and provides information on their distribution and type of growth.

The threat is real. Since bamboos have increased in popularity as garden plants since the 1996 study took place, so too has their frequency increased in the wild in Great Britain and Ireland.

Scientific name	Common name	Distribution/ location	Growth type	Seeding
Arundinaria gigantea	Switch cane / Small cane	One location in Surrey	Running	-
Chimonobambusa marmorea	Marbled bamboo (Kan-chiku)	A persistent relic of cultivation in Guernsey, now exterminated	Running	-
Chimonobambusa quadrangularis	Square-stemmed bamboo	Naturalised in thickets in SW England and Ireland	Running	-
Fargesia murielae	Umbrella bamboo	Persists in neglected parks and estates	Grows in large clumps	-
Fargesia spathacea	Chinese fountain-bamboo	Often grown in gardens	Clump forming	-
Himalayacalamus falconeri	Candy cane bamboo	Persistent relic, Guernsey	Clump forming	-
Himalayacalamus hookerianus	Himalayan blue bamboo	Restricted to milder parts	Clump forming	Sometimes produces seedlings in gardens and greenhouses
Phyllostachys aurea	Golden bamboo	Occasional in neglected parks and estates	Running	-
Phyllostachys bambusoides	Japanese timber bamboo (Madake)	Only in Guernsey	Running	-
Phyllostachys flexuosa	Zig-zag bamboo / Sinuate bamboo (Giuganzhu)	Semi-naturalised in one location in South Wales. A commonly cultivated species but with no other naturalised records	Running	-
Phyllostachys nigra	Black bamboo	Only in one location in each of Jersey and South Wales	Running	-
Phyllostachys viridiglaucescens	Green-glaucous bamboo	Persisted for a few years on a bomb-site in Chelsea, Middlesex, also in Sark	Running	-

Scientific name	Common name	Distribution/ location	Growth type	Seeding
Pleioblastus auricomus	Golden variegated bamboo (kamuro-zasa)	A relic of cultivation	Running	-
Pleioblastus chino	Head-high varie-gated pleioblastus / Maximowicz's bamboo	Recorded in many localities	Running	-
Pleioblastus gramineus	Spiralled grass bamboo	One location in Berkshire	Running	-
Pleioblastus humilis	Toyooka-zasa	A persistent relic of cultivation in Guernsey, now exterminated	Running	-
Pleioblastus pygmaeus	Dwarf bamboo	Fairly often persistent	Running	-
Pleioblastus simonii	Simon bamboo	Rarely persistent	Running	-
Pseudosasa japonica	Arrow bamboo	Long established in numerous localities	Running	Flowers freely and regularly
Sasa palmata	Broad-leaved bamboo	Well established in widely scattered locations	Running	Flowers fairly frequently
Sasa veitchii	Veitch's bamboo	A persistent relic	Running	-
Sasaella ramosa	Hairy bamboo	A relic, sometimes well established	Running	-
Semiarundinaria fastuosa	Narihira bamboo	A persistent relic	Running	-
Thamnocalamus spathiflorus	-	Persists in a few places. Covers large areas in Ireland	Clump forming	-
Yushania anceps	Indian fountain-bamboo, Himalayan bamboo	Established or persistent. Forms thickets	Clump forming but moderately invasive	Sometimes produces seedlings

Table 6a: Species of bamboo found growing in the wild in Great Britain and Ireland up to 1996. (Source: Ryves, T.B., Clement, E.J. and Foster, M.C. 1996. Alien Grasses of the British Isles. Botanical Society of Britain and Ireland, Durham.)

7   Impacts of unmanaged bamboo growth

Many bamboos, and running bamboos in particular, can have significant unwanted impacts on the environment around them. These problems can include:

1)   **Shading**. Some species of bamboo can grow extremely tall. For example, a popular species such as black bamboo (*Phyllostachys nigra*) can reach heights of eight or nine metres when mature. As culms in some bamboo species can grow taller than two-storey (or even three-storey) dwellings, the shading effect and general dominance in any garden cannot be overstated (Fig. 7.1). It is worth noting that Part 8 of the Anti-social Behaviour Act 2003, which deals with problems of tall hedges, is restricted to evergreen shrubs and trees; hence, as species of grass, bamboos are not covered under this particular piece of legislation.

2)   **Rapid rhizome growth**. The underground rhizome network of a

Figure 7.1 Tall bamboo shading a neighbouring property.

Figure 7.2 *Bamboo rhizomes spreading into a lawn from the neighbouring property.*

number of bamboos, such as the *Phyllostachys* genus, can spread distances of over a metre in potentially any direction in a year. It is not unknown to find rhizomes five to six metres long. Most gardens will not benefit from this rate of spread. Lawns have been ruined by the constant emergence of bamboo culms (Fig. 7.2) and it is not uncommon to find bamboos spreading either the entire width of domestic gardens or affecting several gardens at once (see point 4 below).

3) **Structural damage**. Bamboo can damage the built environment by lifting paving and distorting or penetrating soft or decaying tarmac (Fig. 7.3). It can even find its way into a building if a point of access is available. Bamboos have been found growing through stone retaining walls, as well as damaging the concrete bases of sheds (Fig. 7.4), drives, footpaths, pavements, patios (Fig. 7.5), and even roads.

4) **Neighbour disputes**. Bamboo can be the cause of bitter disputes

Figure 7.3 *Bamboo rhizomes interacting destructively with a tarmac pavement.*

Figure 7.4 *A bamboo rhizome discovered penetrating a block of concrete.*

between neighbours. Shading and spreading rhizomes are commonplace, and such invasion can quickly become a source of annoyance. Fear of damage to property can escalate disputes and legal action can be the result.

5)   **Leaf litter**. As with most plants, the dead leaves of bamboos can accumulate (Fig. 7.6). In most bamboo species, the decay of the leaves is very slow, allowing their leaf litter to blanket out other plant growth. In the case of tall species, leaf litter can cause blockages by collecting in gutters and drains.

The advice of the authors is to avoid running bamboos as a choice of plant in a domestic garden. Even some clumping bamboos can create issues if not properly positioned and maintained. The popular scenario of 'planting a bamboo screen' should be avoided. Bamboo (as is the case with any rhizomatous species) is not a hedging plant. If it is placed along a boundary it will typically spread into the neighbouring property, unless very carefully contained, monitored and controlled. Even clumping bamboos spread over time and can encroach across boundaries if planted too close to them and not managed correctly.

Figure 7.5 Bamboo growth lifting paving slabs.

Figure 7.6 Litter from fallen leaves and sheaths.

One solution is to grow bamboos in containers, but even this can require continual monitoring and maintenance, as the rhizomes of running bamboos will make constant attempts to escape. Vigilance is required to ensure the plant does not establish beyond the confines of the container (see the 'Containment' section in Chapter 11).

Bamboos can have a detrimental impact on the built environment. As an extreme example, the authors have encountered a cautionary tale. Mature bamboo in a domestic garden was located close to a boundary in a raised brick planter. The top of the planter was at the same level as the ground in the adjacent property. The bamboo had become fully established over the course of a few years and the rhizomes sought to expand their length in order to support the height of the culms. The wall of the brick planter prevented expansion into the garden and so the rhizomes spread in the other direction, across the boundary line where there was no wall to impede them. The neighbouring house was only 1.5 metres away from the boundary. The rhizomes found ingress into the foundations of the house through service ducting. Once inside, the rhizomes spread and branched beneath the floor, seeking sources of light from gaps where flooring did not sit flush against the skirting. Once they found these sources, they sent up shoots towards them.

Eventually, bamboo began emerging in the corners of almost every room on the ground floor. At its furthest point, healthy bamboo culms were emerging in the corner of the lounge a full 12 m distant from the boundary fence. The occupants of the house had no choice but to evacuate for three months while the entire flooring was pulled up and the bamboo rhizomes removed. This was hugely expensive as well as creating massive upheaval and inconvenience.

Homeowners, surveyors, landscapers and, most of all, bamboo owners need to be vigilant and aware of the impact bamboos can have if left uncontrolled. Case studies illustrating a few of these issues (both structural and amenity) are further described in Chapter 9.

Figure 7.7 Bamboo emerging in the front room of a house.

8 Legislation

At the time of writing, in Great Britain and Ireland there is no specific legislation to cover the sale, planting or spread of bamboo into adjoining properties or the wider environment. In England and Wales, there are no bamboo species included within Schedule 9 of the Wildlife and Countryside Act 1981 (as amended). This schedule lists those invasive non-native plant species which it is an offence to cause to spread into the wild. It is worth noting that Schedule 9 is subject to regular updates and it is possible some of the more problematic species of bamboo could find themselves listed at some point in the future. In Scotland, the Wildlife and Natural Environment (Scotland) Act 2011 states that 'any person who plants, or otherwise causes to grow, any plant in the wild

Figure 8.1 *Bamboo shading house and paved terrace of neighbouring property.*

Figure 8.2 Bamboo causing a significant nuisance by pushing the render away from the boundary wall as well as emerging through the tarmac on the neighbour's side.

at a place outwith its native range is guilty of an offence'. This would include bamboo species, since Scotland is not their native range, and ensures that a rapid response to new populations could be undertaken. Such measures, however, would be limited to the wild.

Shading (Fig. 8.1) is not covered by Part 8 of the Anti-social Behaviour Act 2003, which deals with high hedges of evergreen shrubs and trees. However, local authorities and the police do have power under the Anti-social Behaviour, Crime and Policing Act 2014 to issue Community Protection Notices (CPNs) if they are satisfied the individual against whom the order is being issued has persistently, or continually, acted in a way that has had a detrimental effect on the quality of life of those in the locality. However, local authorities or the police are unlikely to instigate action under the legislation if only one neighbour is being affected by the nuisance. The Act is tailored towards a persistent series of complaints that have not been satisfactorily addressed by the individual and where the nuisance is affecting several parties in the neighbourhood. The implementation of the legislation is left to the interpretation of individual local authorities and the police, so different responses are likely to be given by different councils and local Wildlife and Crime Officers.

Common nuisance law would apply to bamboos if the owner allowed their bamboo to cause a problem to their neighbours, preventing the 'quiet enjoyment' of their property (Fig. 8.2). This is the same principle as in Japanese knotweed disputes between neighbours and the damage and cost of remediation could prove similar. It is therefore advisable that bamboo owners consider carefully the impact their bamboo can have on others and take appropriate steps to minimise such effects (Fig 8.3).

This kind of legal action can prove to be something of a grey area and the success or otherwise of such action will vary greatly from case to case. Consequently, landowners and managers should consider seeking appropriate legal advice before taking any action. As in all things, it is always better for property owners to talk to each other in the first instance and establish a mutually agreeable solution before rushing to court. Legal action can be a protracted and expensive affair and the only real winners are usually the legal firms.

Figure 8.3 *Bamboo rhizomes in a raised bed are destabilising the boundary wall and pushing it over.*

9

Case Studies

Case Study 1

Figure 9.1 *Showy yellow-groove bamboo to the left, black bamboo to the right, either side of the tree. Both bamboos had been planted next to the fence.*

Site survey

Two species of bamboo had been planted as part of a screening hedge in a residential garden in West Sussex. Although planted next to the fence line, no root barrier or any other form of containment was implemented. Both species chosen by the householder – *Phyllostachys aureosulcata* 'Spectabilis' (showy yellow-groove bamboo) and *Phyllostachys nigra* (black bamboo) – are known for both their aggressive rate of spread and their very tall culms (Fig. 9.1). Despite this, the nursery from where the plants were purchased did not indicate the bamboo would create any issues. In 2018, the owner of the neighbouring property dug out rhizome encroachment of over a metre into their property (Figs. 9.2 & 9.3). In 2019, further rhizome removal occurred and in 2020 some paving slabs in the patio close to the neighbour's house were seen to be raised. On lifting the slabs, thick bamboo rhizomes from the *Phyllostachys nigra* were evident, with the closest being only 60 cm from the side of the house (Fig. 9.4). A rhizome of 1.7 m in length was measured under the patio. At this time, the *Phyllostachys nigra* was over four metres tall and the *Phyllostachys aureosulcata* 'Spectabilis' had reached a height of around 5.5 m.

Figure 9.2 *Encroachment beneath the boundary fence from the black bamboo. A new culm shoot can be seen to the left (yellow arrow). This image was taken in May 2019. The encroaching rhizomes had previously been cut off along the boundary line in November 2018 (some of the cut edges can be seen beneath the fence).*

Impact of the bamboo

The height of the bamboo was causing some shading to the neighbouring property, though the neighbours were more concerned about the bamboo culms emerging in their shrub bed, lawn and patio. They were particularly concerned about the risk of damage to the house itself. The owners of the

Figure 9.3 *New rhizomes (red arrow) and thin immature culms (yellow arrow) emerging from the showy yellow-groove bamboo in May 2019, following full rhizome removal up to the fence line in November 2018.*

Figure 9.4 *Rhizomes from the black bamboo discovered beneath the paved area in June 2020.*

Figure 9.5 Rhizome pushing through the cement beneath the paving slabs.

Figure 9.6 Dense mass of bamboo rhizomes beneath the black bamboo revealed during excavation.

bamboo were unclear as to the extent of the problem, since they had seen no such encroachment from the bamboo within their own property and had received assurances from the nursery which had sold the bamboos that there was nothing to be concerned about. Once the rhizomes were discovered beneath the patio (Fig. 9.5), the owners of the bamboo saw for themselves the seriousness of the situation and agreed to remedial action. A site survey revealed that the ground sloped gently down from the bamboo towards the neighbouring property. Since rhizomes will typically find the path of least resistance in their search for moisture and nutrients and will often grow downhill, this slope helped explain why the neighbours had been bothered by spreading rhizomes while the owners of the bamboo had seen no such spread occurring on their property.

Remedial action proposed

Containment, though feasible, would have been impractical at this stage due to the high cost. In addition, the closeness of both bamboos to the boundary fence would have meant any containment measures would have had to be implemented on the neighbours' side. The owners of the bamboo agreed that the best course of action would be to completely remove both bamboos.

Period and nature of works implemented

In 2020, the bamboos were cut down and all associated rhizomes in both properties were physically removed from the ground and disposed of off-site (Fig. 9.6). It was not possible to remove rhizomes where they were entangled with the roots of surrounding shrubs. Where this occurred, rhizomes were cut off and left *in situ*. A monitoring programme has been implemented to ensure any growth that emerges from these remaining rhizomatous fragments will be treated with herbicide. This programme is to run until 2026.

Lessons learned

No matter how much you may like the look of bamboo, or what those selling it to you may say, bamboo comes with a responsibility to control, maintain and contain it to prevent it causing a nuisance to surrounding properties. Planting bamboo, particularly a tall running bamboo, immediately next to a boundary line is highly inadvisable. Topography can have a bearing on the direction bamboo spreads. Sometimes, when an issue arises from a neighbouring source, the best way to get things resolved is to allow owners of the bamboo to see the extent of the problem with their own eyes. In this particular case, the shock of how far the rhizomes had spread beneath the paving prompted the owners of the bamboo to agree to the complete removal of the plants they loved.

Case Study 2

Figure 9.7 *Mature bamboo growth, July 2013.*

Site survey

Bamboo had been planted as a screen in the garden of a house in Cambridgeshire (Fig. 9.7). It was planted along a garden boundary with no measures for containment or subsequent management. Consequently, the bamboo spread into the adjoining garden, where it was affecting decking and paving slabs (Fig. 9.8). Bamboo had also spread into the front garden of the original bamboo owner's property and damaged a weed-suppressing membrane. At the time of the authors' survey, the bamboo had attained a height of four to five metres. Relations between the neighbouring property owners had become strained.

Impact of the bamboo

The height of the bamboo (up to three metres taller than the two-metre-tall boundary fence) was causing significant shading at various times of the day in the neighbouring property. Minimal damage had been caused to the neighbours' decking and paved pathway, though concerns were high over

Figure 9.8 Bamboo lifting paving slabs and causing a trip hazard.

Figure 9.9 Bamboo emerging in a driveway. New culms have sharp points and edges which could prove damaging to people and vehicles.

how much damage the bamboo would continue to cause if left unabated. Both properties had emergences of bamboo culms in their driveways and concerns were expressed as to the sharpness of the emergent culms and the threat they posed to pedestrians and vehicles using the driveways (Fig. 9.9).

Remedial action proposed

Containment would have been impractical at this stage, so a full herbicide treatment programme using a glyphosate-based herbicide to rid both properties of the bamboo was proposed.

Figure 9.10 *Bamboo following initial foliar treatment.*

Period and nature of works implemented

Works began in July 2013 and completion was reached in 2018. The bamboos in both affected properties were subject to a herbicide treatment programme (Fig. 9.10). Works commenced with foliar spray treatment and included the cutting down of all culms in the winter of 2013/14, when stumps were injected using a needle-free technique pioneered by Brian Taylor in early 2014 (Fig. 9.11). Thereafter, all properties were inspected and any regrowth was spot-treated.

Figure 9.11 Brian Taylor conducting a winter stem injection using his needle-free method.

Lessons learned

The use of herbicide to effectively control bamboo requires protracted treatment and typically takes longer than people think. Treatment times and frequency can be greatly increased when rhizomes are growing beneath surface obstacles (e.g. slabs, decking, tarmac, landscape membrane, etc.). Disturbance of bamboo material (e.g. being driven over) can also lead to an extension of the treatment programme. Winter works (i.e. cutting down mature culms) generate large volumes of plant debris that need to be safely disposed of. Bamboos, particularly running bamboos, are not suitable for planting as a hedge or screen close to a property boundary unless efficiently contained.

Case Study 3

Figure 9.12 *Bamboo dominating the front garden of the townhouse.*

Site survey

Two bamboo plants were planted in the front garden of an end-of-terrace townhouse in Hampshire. The houseowners purchased the bamboo from a reputable garden nursery, who informed them the bamboo was 'non-invasive'. Over the course of the next 10 to 15 years, the bamboo grew to completely dominate the front garden (Fig. 9.12), so that the homeowners were frequently cutting back and treating culms and rhizomes with herbicide. At the time of the authors' survey, the bamboo was around six metres in height. Fortunately, the bamboo seemed to be contained within the boundary walls of the property and there was no visible evidence of spread into adjoining areas, with the exception of a single rhizome by the front gate.

Impact of the bamboo

The bamboo had exceeded the height of the guttering at the base of the roof and was causing considerable shading to the dwelling. Rhizomes were

Figure 9.13 Heaving of the path caused by bamboo rhizomes. New bamboo shoots can be seen breaking the surface of the gravel (arrowed). The houseowners had been cutting back culms and applying herbicide to visible rhizomes in this area.

Figure 9.14 Bamboo rhizome beneath the path immediately adjacent to the step leading to the front door.

Figure 9.15 *The thick mass of bamboo rhizome revealed by excavation.*

evident throughout the front garden and the client was struggling to keep the majority of the garden clear of culms. There was evidence of heave in the path leading from the road to the front door (Fig. 9.13). One rhizome had escaped from within the boundary of the property through the front gate area and was lifting the tarmac in the footpath outside (Fig. 9.14).

Remedial action proposed

The client wanted the bamboo to be removed so that the front garden could be redeveloped. Accordingly, the excavation of rhizomes and the off-site disposal of all bamboo material were proposed.

Period and nature of works implemented

The bamboo was treated with a foliar application of a glyphosate-based herbicide a few weeks prior to commencement of removal works in order to weaken the plants. Excavation began in September 2018 and took three days to complete. The excavation revealed a dense network of rhizomes covering the entirety of the front garden to a depth of around 17 cm (Fig. 9.15). Despite the shallow depth, the thickness of the rhizomes and the density of the rhizome mass meant that the use of a mechanical mini-excavator was

Figure 9.16 *Bamboo rhizome growing along the exterior wall of the house. Penetration into the wall has occurred just to the right of centre of this photo.*

essential. Rhizome penetration into the front wall of the house below the ground level was discovered, though it was not clear if the bamboo rhizome had created the point of ingress or whether the mortar in the brickwork was previously weakened at that location (Fig. 9.16). A dense mass of rhizome material was discovered in the two places where the original bamboos had been planted. These dense masses were so tightly bound to the walls – one was the front boundary wall and the other was the front exterior wall of the house – they had to be left *in situ* for fear of damaging the walls. Some of this material was subsequently removed using a disc cutter.

On completion of the works, full eradication had not been possible to achieve. Some rhizomes remained *in situ*, close to the foundations of boundary and residence walls, beneath the tarmac in the public footpath outside and in the brick wall of the house where rhizome had penetrated through into the building. All of these rhizome fragments were treated with herbicide on their cut edges. The clients were able to redevelop their garden, provided they left a narrow corridor around the edge of the property untouched. This corridor has been subject to an ongoing monitoring programme, whereby any regrowth is identified and treated. After a full 20 months following excavation works, a flush of new growth emerged from a truncated section of rhizome beneath a boundary wall (Fig. 9.17), highlighting the importance of constant monitoring.

Lessons learned

Just because people at garden centres say a plant is 'non-invasive' does not mean they are correct. A little research prior to purchase can go a long way. Bamboo rhizomes of many species do not easily snap and the density of the rhizome network makes excavation extremely labour-intensive, even if using a mechanical digger. Rhizomes can penetrate the wall of a house if the mortar is weakened or if there is already a point of entry. If there is the slightest risk of rhizome fragments remaining in the ground, post-removal monitoring is crucial, with follow-up treatment as necessary. Fragments of bamboo rhizome can retreat into a dormant state with no obvious evidence of activity for several months, perhaps years.

Figure 9.17 *Healthy growth emerging from a truncated section of rhizome, which had displayed a period of inactivity lasting over a year. Whenever such fragments of rhizome remain following excavation, post-removal monitoring is essential.*

Case Study 4

Figure 9.18 *A dense, tall area of bamboo affecting the amenity and usability of one of the gardens.*

Site survey

A housing association owns six detached residential properties with large rear gardens in a semi-rural location in Hampshire. Bamboo was present at the bottom of the gardens and could be found in five of the six properties (Fig. 9.18). The details surrounding the planting of the bamboo were unknown, but it is probable it was planted sometime in the late 1990s or early 2000s. At the time of the authors' survey, the bamboo was covering an area of almost 1000 m² and was beginning to spread into adjoining woodland beyond the ends of the gardens. The bamboo across the several properties ranged from between five to seven metres in height.

Impact of the bamboo

Although the bamboo could be found in the furthest third of each affected garden, encroachment further into the gardens and towards the houses was occurring. In two gardens, a fence had been erected to segregate the bamboo area. In one of the properties, the resident was cutting back any bamboo that

Figure 9.19 *One of the residents has tried to fence off the bamboo, but is still fighting a battle with all the culms that emerge inside the fence and encroach into the lawn.*

emerged beyond the fence; in the other, the attempt at containment had failed and culms were emerging in the lawn (Figs. 9.19 & 20). Environmentally, the woodland would suffer a significant impact if the bamboo was allowed to continue its spread and become established.

Remedial action proposed

The bamboo was too well entrenched to be merely contained and, as two of the residents had discovered, fencing the bamboo off did not halt its aggressive growth-habit. The clients were aware it was a problem that required decisive action, but they felt unable to find reliable advice and an appropriately costed solution. A full treatment programme was recommended, with the aim of ridding all the gardens of the bamboo, and removing the risk of encroachment into the adjoining woodland.

Figure 9.20 *Bamboo encroaching into the lawn of another of the properties. The fence to segregate the bamboo from the rest of the garden has failed. This picture gives an indication of how far rhizomes can extend from the main bamboo plants.*

Period and nature of works implemented

Works began in March 2017 and are ongoing at the time of writing. The bamboo in all affected properties is under a herbicide-treatment programme, which will take several years to complete. The culms that have emerged in the edge of the woodland have been included as part of the programme. Works commenced with foliar-spray treatment (Fig. 9.21), leading to the cutting down of all culms in the winter of 2017/18. The cut culms were shredded and reused on site as mulch. Stumps were injected with herbicide at this time. Thereafter, all properties are being inspected and any regrowth (Fig. 9.22) spot-treated.

Figure 9.21 Jim Glaister applying a foliar treatment to bamboo using a knapsack sprayer with an extended lance.

Lessons learned

If left unmanaged, bamboos will quickly come to dominate and to encroach into surrounding areas. The longer bamboo is left, the more effort will be required to control or remove it.

Figure 9.22 Regrowth recorded in the third year of the treatment programme.

10 Common invasive bamboo species in Great Britain and Ireland

There are many bamboo species that can potentially cause problems in Great Britain and Ireland. Those more commonly favoured by landscapers and gardeners are (Figs. 10.1-6):

1) Golden Bamboo (*Phyllostachys aurea*);
2) Black Bamboo (*Phyllostachys nigra*);
3) Broad-leaved bamboo (*Sasa palmata*);
4) Showy yellow-groove bamboo *(Phyllostachys aureosulcata* 'Spectabilis');
5) Green bamboo / David Bisset bamboo (*Phyllostachys bissetii*);
6) Umbrella (or Pingwu) Bamboo (*Fargesia robusta* 'Pingwu').

It is not uncommon to find more than one of these species planted on the same site. Not all bamboo species should be considered invasive, though any bamboo has the potential to spread and dominate over time. The first five of the six species listed above have a well-established history of becoming problematic, as they spread quickly and require vigilance and active containment. The sixth species on the list is often promoted as non-invasive, due to its clumping habit, but any species capable of reaching a height of up to five metres is capable of spreading across a boundary if not properly located, contained and maintained. The authors therefore advise against planting any of these species in an enclosed location, particularly a domestic garden. Table 10a lists the bamboo species most commonly planted in gardens in Great Britain and Ireland.

Bamboo species	Brief description	Average height range (m)	Time to full height (years)
Chimonobambusa marmorea	Low-growing (to 2 m high); young shoots lime-green marbled brown and white; purple-flushed adult culms; grass-green leaves.	2 - 3	2 - 5
Fargesia murielae	Clump-forming; yellow-green culms; bright green leaves.	2.4 - 4	5 - 10
Fargesia nitida 'Nymphenburg'	Compact; clumping; slender purple-flushed green culms; narrow leaves.	2.5 - 4	5 - 10
Fargesia robusta 'Pingwu'	Upright and vigorous; leaves slightly larger than other *Fargesia* bamboos; young culms yellow and red; sheaths fade to almost white.	4 - 5	5
Phyllostachys aurea	Vigorous; upright; running rhizomes; bright green culms that turn soft yellow in very strong light and with age.	5 - 6	5 - 10
Phyllostachys aureosulcata 'Spectabilis'	Vigorous; running rhizomes; yellow culms with green sulcus; distinguished by occasional zig-zagging of the lower part of culms.	6 - 14	5 - 10
Phyllostachys bissetii	Vigorous; mainly upright; green culms, occasionally with purple flush, fading to yellow-green with age; rich green leaves; initially forms dense clumps, with rhizomes running with age.	4 - 6.5	5 - 10
Phyllostachys nigra	Large; running rhizomes; culms green at first, becoming blackish-brown with age: narrow leaves.	2.5 - 8	5 - 10

Bamboo species	Brief description	Average height range (m)	Time to full height (years)
Phyllostachys viridiglaucescens	Very vigorous; running rhizomes; green culms maturing to yellow-green; leaves bright green, glaucous beneath.	10	5 - 10
Pleioblastus hindsii	Vigorous; clump-forming; dense thickets of olive-green culms; blue-green leaves.	2.5 - 4	5 - 10
Pleioblastus viridistriatus	Compact dwarf bamboo; upright; slow growing; non-flowering; green culms turn purple with age; long narrow leaves, golden with distinct vertical green striping.	1 - 1.5	5 - 10
Pseudosasa japonica	Vigorous; spreading; olive-green culms; copious narrow, shiny green leaves.	2.5 - 8	5 - 10
Sasa kurilensis 'Shima-shimofuri'	Spreading; slender culms; leaves variable, with creamy-white stripes.	0.9 - 3	5 - 10
Sasa palmata	Vigorous; lower-growing; dark green leaves, with tips and margins turning pale brown in winter, giving a variegated effect.	1.5 - 3	5 - 10
Sasa veitchii	Vigorous; dense-growing; spreading habit; dark green leaves, withering at the margins as winter approaches, giving a variegated effect.	0.5 - 1	5 - 10
Shibataea kumasaca	Dwarf, compact species; dense, leafy clumps; pale green culms, turning dull brown with age; dark green leaves.	0.5 - 1	5 - 10

Table 10a Examples of bamboo species most commonly found in gardens in Great Britain and Ireland.

Figure 10.1 Golden bamboo.

Figure 10.2 Black bamboo.

Figure 10.3 Broad-leaved bamboo.

Figure 10.4 Showy yellow-groove bamboo.

Figure 10.5 Green bamboo / David Bisset bamboo.

Figure 10.6 Umbrella (or Pingwu) bamboo.

11 Developing a Management Strategy - practical advice for managing, controlling or eradicating bamboo

Dealing with an infestation of bamboo needs a Management Plan, which would contain the various stages of survey and assessment. It would include a review of current management and provide applicable recommendations, as well as monitoring post-treatment and, where appropriate, revegetation.

Training in the identification, surveying, assessment and management of invasive non-native plants is available from organisations such as the Property Care Association (PCA) and BASIS.

Surveying

The survey is the first step in developing a Management Plan and should seek to answer as many of the following questions as is practicable:

- What is the site history?
- What are the owner's details and the contact details for any other affected properties?
- How did the bamboo get to site?
- What is/are the species of bamboo?
- When and where was the bamboo planted?
- What steps (if any) were taken to contain the bamboo and limit the rhizome spread?
- Have there been any previous treatments or other control attempts and, if so, what are the details?

Additionally, the surveyor will need to:

- Make sure the bamboo is correctly identified – note there may be more than one species present.
- Identify the location(s) of the bamboo plant(s) and produce a scale plan showing rhizome extent (where known or identified by making trial digs).
- Check access for personnel, machinery and bamboo removal routes (as required).
- Determine any existing damage to hard surfaces / structures.
- Determine any existing damage or loss of amenity use to garden / adjoining gardens (as appropriate).
- Identify any spread to the local environment.
- Check for any complicating factors, such as the proximity of water courses or underground services.

The most difficult part of any survey will be identifying the bamboo species present. With so many bamboo species in existence, and with the difference between many of them being so slight, it is not uncommon for bamboos to be identified by surveyors at the genus level but not necessarily at the species or family level. Be aware that more than one species of bamboo can be present at the same site, but not all of them may be causing a problem.

Integrated management

As outlined in the following sections of this chapter, there are several approaches available for the control or eradication of bamboo. The professional contractor should evaluate each site and choose whichever methods are more suitable for the site and the client's needs that would be both safe and effective. It may be that more than one method will be utilised.

The purpose of Integrated Weed Management is to use different control practices (perhaps in combination) to ultimately use less herbicide and make use of more environmentally sustainable techniques where feasible. The five main procedures that can be included within Integrated Weed Management are preventative, mechanical, cultural, biological and chemical.

Preventative measures include not planting running bamboos in unsuitable locations in the first place and not allowing them to cross boundaries, establish in the wild or in other unsuitable areas.

Mechanical methods include the physical excavation of the bamboo and its rhizomes, removing all excavated matter to a licensed landfill or other suitable disposal site.

Cultural involves the containment of bamboo and would include ongoing maintenance.

Biological control methods comprise the use of natural biological agents (e.g. insects or fungi) that control the growth of the host plant. The authors are not aware of any biological control agents available for bamboo in Great Britain and Ireland at the time of writing.

Chemical would see the bamboo being controlled by the use of herbicides.

It should be possible to develop an Integrated Management Plan for bamboo from the available means. For example, pre-applications of herbicide to the bamboo prior to an excavation will reduce the viability of the excavated rhizome and may also reduce the scope of the necessary excavation.

For a non-excavation strategy, incorporating cultural control methods as part of a herbicide programme (by severing and removing long-running rhizomes for example) may well reduce the impact and the use of herbicides on a site.

Control and removal methods

If the aim is to retain bamboo but mitigate the risks and limit the spread of the plant, please see the 'Containment' section below. If the aim is to get rid of the bamboo altogether, there are two principal methods available: herbicide treatment and excavation.

Herbicide control of bamboo

Bamboo is a grass. A very large grass admittedly, but it is a monocotyledon and part of the grass family. As such, the most suitable choice of herbicide would normally be a glyphosate-based product. Please note that all herbicides should be used in strict accordance with the recommendations on the product label. The adding of non-approved products or the mixing of different herbicide products (unless specifically approved on the product label) may be highly dangerous and even illegal. Such practices will not improve the efficacy of the product and may actually reduce the effectiveness.

Figure 11.1 *Treating bamboo with herbicide using a knapsack sprayer and a long lance.*

The end user is responsible for the safe and correct use of herbicides. The correct dose/dilution for the herbicide in use, as well as any specific approvals/restrictions on use, will be found by reading the specific product label. Appropriate personal protective equipment (PPE) should be used when mixing and applying herbicides. Advice on what PPE is appropriate for the herbicide in question will be provided within the product label.

The purpose of a herbicide treatment programme is to kill off the rhizome of the bamboo. With a clumping bamboo, this goal will be achieved faster than with a running bamboo. Programmes will typically take <u>at least</u> three to four years to complete on healthy bamboo and can sometimes take longer, particularly for larger colonies. The longer the rhizomes, and the larger the area the bamboo covers, the longer the treatment will take to achieve completion. Table 11a details a typical herbicide programme.

Year 1	Summer to early autumn: Foliar herbicide treatment (either one or two applications, depending on size of infestation).
Year 2	Winter: Cutting of culms and stem injection (see details below).
	Summer to early autumn: One or two visits during the summer/early autumn months to spot treat (foliar application or weed wipe).
Year 3	Mid to late summer: Single visit to spot treat any re-growth (foliar application or weed wipe). Check close-to-surface rhizomes with hand excavation. If grey/brown, consider weed wiping these.
Year 4 (and subsequently)	Mid to late summer: Single visit to spot treat any re-growth (foliar application or weed wipe). Check close-to-surface rhizomes with hand excavation. If grey/brown, consider weed wiping these.
	Consider the programme completed when there has been no new growth for two consecutive years and if only grey/brown rhizomes can be found when manual excavation in the bamboo area takes place.

Table 11a: *Example of a typical herbicide-treatment programme for bamboo*

Figure 11.2 *Bamboo rhizomes can be long and do not easily break.*

There are three principal methods of applying herbicide to the leaves, stems and, where exposed, rhizomes of the target plant: foliar application (in which herbicide is sprayed directly onto the plants), weed wiping (a more targeted hand application) and stem injection (using a specialist applicator). All three of these methods are available for treating bamboo if using a suitable glyphosate-based product.

Foliar application: For best results, carry out treatments between summer and early autumn (July-September). Early spring applications, when new culms are emerging, are unlikely to be successful. At this time, the bamboo is moving sucrose up into the new growth from the rhizomes. A glyphosate-based herbicide is more likely to be effective when it is drawn down into the rhizomes and this is more likely to occur once the culms have reached their optimum height. Since it is an evergreen, bamboo can be treated during the winter months, but the summer to early autumn period is when treatments will be the most effective. Given the height of many bamboos, a long extending lance may be required to apply the herbicide to the leaves (Fig. 11.1).

Weed wiping: This technique is more suitable for small infestations, for follow-up visits to treat new regrowth, or for when the bamboo is growing in areas where it is surrounded by plants or grass which the property owner wishes to preserve. Essentially, this method involves applying a glyphosate-based herbicide directly to any green stem or leaf by contact wiping, using either a specialist wiper or a sponge. Frequently in years two to four of a herbicide programme there may be little or no leaf cover and few if any live culms visible. However, the shallow rhizomes can be uncovered in the soil and, if still retaining colour, may be treated effectively using this technique.

Stem injection: Unlike other hollow-stemmed plants (e.g. Japanese knotweed), bamboo culms are too hard to inject using needles. However, a cut-and-fill injection method can be employed instead. During the winter months (October to March), bamboo culms can be cut down to just below the second node from the soil surface, whereby the remaining culm stump can be injected from above into the hollow stem. It is important to ensure the correct dose is applied to each stump and that the bamboo is not overdosed. One means of achieving this is to use a specialist injector gun or a syringe (without the needle). Where a large number of stumps are to be treated, it is advisable to add a blue tracer dye to the herbicide to ensure no stumps are accidentally missed or double-dosed. A correctly diluted herbicide is more effective and more readily absorbed than an undiluted product. Using undiluted herbicide is more likely to exceed the product label recommendations (and therefore break the law) and overdose the plant (which could induce temporary dormancy). Always refer to the recommendations given on the product label to determine the correct dose.

Excavation of bamboo areas

Bamboo rhizomes are woody and very strong. They are difficult to cut and, in most species, do not snap when pulled from the ground (Fig. 11.2). Rhizomes have a tendency to branch, which can anchor them into the ground and make them harder to pull out of the soil (or a structure). The rhizome mass of an established bamboo plant is densely packed and excavating bamboo can be an extremely labour-intensive affair. If feasible, it is always best to use a mechanical excavator to remove bamboo rhizomes. Be aware that, whatever the means of extraction, removal is likely to take longer than anticipated.

The one advantage bamboo has for contractors removing it from the ground is that the rhizomes are generally shallow. It is unusual to discover bamboo rhizomes deeper than half a metre, and it is more typical to find them in the top 30 cm of soil. Yet the roots of individual plants can be found at a greater depth, typically up to a metre deep.

Even with a mechanical digger, removal from the ground can prove to be a struggle due to the nature of the rhizome's structure. Sometimes, bamboo rhizomes can be separated easily from the ground (so reducing the quantity of material to be removed from the site), though this should not be relied upon as an outcome. Running rhizomes can be long and often entangled with each other, so the simple act of removal can be difficult. If the bamboo is growing in a restricted area, such as a brick planter, the base of the bamboo and the associated rhizomes can be so densely packed that removing them can be akin to digging out reinforced concrete (Fig. 11.3).

Figure 11.3 Bamboo rhizomes can prove difficult to excavate.

Rhizomes, being regenerative, should always be properly destroyed or disposed of. However, there is some flexibility in what to do with the culms once they have been cut down. If space permits, they can be shredded and mulched on site. Incineration of culms, though effective, carries some safety concerns and is not recommended. See 'Disposal of bamboo culms' below for more information and options for disposal.

Excavating work must consider the full extent of a bamboo's rhizomes. They can and do grow into walls and other structures, as well as under hard surfaces. Such material cannot be easily removed. In these circumstances, once excavation has extracted all that it is possible to take, it is advisable to apply a glyphosate-based herbicide to any uncovered rhizomes that remain and to cut edges. A programme of monitoring for subsequent regrowth should be implemented, involving the treatment of regrowth with herbicide. Monitoring and treatment should also be implemented for any neighbouring properties where the rhizomes have encroached but cannot be removed. Any fragments of rhizome remaining *in situ* on a neighbouring property will be likely to regenerate new growth and spread back onto the property from where it has been excavated. A root barrier could be installed along the boundary to prevent such future encroachment, but this is not a fool-proof method – see 'Containment' section below.

Excavation, by its nature, is intrusive and disruptive to the general amenity of the site. On completion of the works there will be a hole to fill and the need to replace or repair hard surfaces, shrubs, hedges, lawns and/or fences.

Disposal of bamboo culms

Whether treating bamboo under a herbicide programme or fully excavating the bamboo, the culms will need to be cut down at some point (Fig. 11.4). Culms are tough and not easy to cut, so a good pair of loppers will be required. As mentioned above, culms can be shredded with a mechanical shredder and used as mulch. Alternatively, they can be dried (whereby they become referred to as canes) and repurposed in a number of ways:

Figure 11.4 Cutting down bamboo culms.

for example, for horticultural use, to create a screen wall or as a building material. Local allotment societies may be glad to receive a donation of dried bamboo canes. Schools could also make use of dead canes, as they can use them to create bug hotels to capture the interests of the children.

If cut culms are to be removed immediately for disposal off-site, there are currently no restrictions in place as to where they can be taken. Unlike, say, Japanese knotweed, bamboo is not a controlled waste and can be disposed of at landfills and amenity tips alike as green waste.

It is a tempting idea to burn bamboo culms, though there is an inherent danger with this approach that can take the unwary by surprise. The culms, even if dried out as canes, often contain water in the internode sections. When heated, the water turns to steam and the cane explodes. The sharp, burning fragments can travel considerable distances at high speed.

Post-treatment and post-excavation monitoring

When stressed, bamboo can retreat into a dormant state for a period of time. This has been observed in rhizome fragments that remain *in situ* following an excavation and, more commonly, in bamboo that has been treated with herbicide.

The use of a glyphosate-based herbicide in particular can induce a period of apparent inactivity. Glyphosate works by being absorbed through the foliage and green stems and transported (or translocated) within the internal circulatory system of the plant down into the roots or rhizomes. The glyphosate is moved within the plant by the same modes of action as sucrose. In the spring, when bamboo rhizomes are forming new buds, sucrose is drawn to them, and the glyphosate travels along with it. If there is enough glyphosate present in the rhizomes, the buds will be poisoned by the chemical and will die. Yet, unless the rhizome itself dies, this process will be repeated until the levels of glyphosate within the rhizome are reduced to a low enough state to permit new buds to develop and grow. It may take several years for the glyphosate to be fully exhausted. The seeming inactivity observed above ground caused by the glyphosate in the rhizomes is best described as a period of pseudo-dormancy, as opposed to a true dormant state. Low levels of glyphosate within the plant can also distort new growth, creating emerging bamboo that is not immediately recognisable (usually referred to as 'bonsai' regrowth).

Whether caused by herbicide or by stressed rhizome fragments, the fact that some or all of a bamboo plant can become inactive for a season or longer makes it essential to build a period of monitoring (minimum two years) onto the back end of any course of bamboo remediation.

Revegetation

The removal or treatment of bamboo will leave an empty space that will be open to colonisation by other plants, including many invasive species with airborne seeds. A bamboo management plan should include a recommendation for revegetation. Ideally, the bamboo area should be re-planted with desirable species as soon as is feasible. In the case of full excavation, revegetation can be carried out when removal works have been completed.

With herbicide treatment, the process of revegetation is more complicated. By year two of the programme, the bamboo's regrowth will be sparse and quite weak. This will afford significant opportunities for colonisation of the area by undesirable species. These can be treated as part of the continuing bamboo treatment programme, or can be managed with more traditional manual means, so long as viable bamboo rhizome is not removed from the area.

It is possible for replanting to take place after two to three years from commencement of the treatment programme. Nevertheless, it should be borne in mind that the treatment programme may not have been completed by this point and that access to new emergent bamboo growth will be required to allow weed wiping or spot spraying. If at all possible, revegetation should be left until the bamboo treatment programme has been completed and two years of monitoring with no new regrowth has taken place.

Containment

For anyone set on the idea of having bamboo on their property, containment and control (as well as choosing the right species in the first place) are keys to enjoying the plant without all the nuisance that can go with it. However, depending on the species, successful containment does require care. There are several strategies available:

Figure 11.5 *Bamboos in containers (either side of containerised phormium).*

Containers

Containerising bamboos can be effective for some of the smaller clumping species (Fig. 11.5), though the bamboo will still need to be kept under control. Ceramic containers can break if rhizomes grow too large for the space they are confined within, so cutting and thinning (see below) are very important. Running bamboos have a tendency to escape from containers, often by expanding over the top edge of the container. Rhizomes will then make their way to the ground, where they will embed themselves, spread beyond the container and produce culms. Both running and clumping bamboos can find their way through the drainage hole at the base of a container, so be sure to give some thought to the placement of the container to prevent this occurring. As with most control methods, when it comes to containerising bamboo, vigilance is the key. Just because a bamboo is in a container, do not assume it will stay there. Limit the top growth to prevent the need for rhizome expansion. Regularly inspect for escaping rhizomes and cut them off before they reach soil.

Thinning and cutting

Once culms reach their optimum height during a growing season, they will rarely (if ever) increase in size. Instead, the bamboo will simply produce a new crop of culms each year. With new culms growing each season, the bamboo will consist of a combination of old and new culms (see Fig. 6.2). Individual culms can live for over ten years. It is therefore important to thin out the bamboo periodically once it has become established by cutting back the older culms. Imposing and maintaining a maximum height to the bamboo will also help restrict how far rhizomes will spread. The taller the bamboo, the further the rhizomes will need to spread in search of sufficient nutrients to support the height.

Edging

Edging involves cutting back the rhizomes as they expand towards the limits of the designated growing area. With the majority of the rhizomes being close to the surface, edging can be carried out with a good sturdy spade with a reasonably sharp blade. As well as aiding containment, cutting the rhizomes helps keep the bamboo healthy. Bamboo growers recommend carrying out edging twice per year.

Edging is, like thinning, an ongoing process. In order for the rhizomes to be visible, the area around the bamboo would have to be maintained as soil only (no paving, no gravel, no shrubs). Alternatively, a shallow v-shaped trench (approximately 30 cm deep should be sufficient) could be cut and maintained along the outer extent of the bamboo's containment zone. This would be particularly useful between a bamboo and a fence line. Any rhizomes that cross the gap formed by the trench would be clearly visible and could be cut, provided the trench is being regularly monitored

It is probable that edged rhizomes will produce new culms at the cut ends. These would need to be regularly cut back to prevent them obscuring the edging area.

Root barrier

Some nurseries advise their clients to install a specialist root barrier prior to planting bamboos in order to prevent spread into adjoining properties. For this to have any chance of success, several factors need to be taken into account.

First, the barrier needs to be fit for purpose. Bamboo rhizomes grow quickly and can be quite pointed at the tips. A standard barrier will simply tear

Figure 11.6 *Installation of a root barrier along a boundary.*

under the pressure the rhizome will exert – and even concrete can prove ineffective unless installed at a significant thickness, as it will be prone to some crumbling and cracking over time under the exertions of the bamboo rhizomes. Therefore, a stronger and heavier geotextile will be required, such as specially formulated Bamboo Root Barriers.

Second, a root barrier needs to be installed to a depth of between 60 and 120 cm (Fig. 11.6). The trench dug to receive the root barrier needs to be of sufficient length to cover, not only the width of the visible bamboo plant(s), but also to allow for the likely distance the rhizomes will grow – a minimum of three to five metres in each direction beyond the bamboo plant – otherwise bamboo will simply grow around the end or ends of the barrier. The barrier can be installed vertically or on a batter. Battering, or angling, the barrier will encourage rhizomes to grow upwards towards the surface, where they can be seen and more easily dealt with, instead of sending them deeper. Deep rhizomes can cause more issues and be harder to find. Compacting the base of the barrier-receival trench before backfilling can also help discourage rhizomes from going deeper.

Even if a root barrier is correctly installed, bamboo rhizomes – being close to, and often visible along, the surface – can simply grow over the top of it. It is therefore necessary to keep monitoring the barrier and deal with any encroaching rhizomes that are attempting to cross. The condition of the barrier itself should also be checked regularly (at least annually) to ensure it has not been damaged.

Whichever method is used, maintenance is vital – particularly for running bamboos. If a bamboo is left to its own devices it will not take long to completely dominate the area in which it has been planted. In all likelihood it will also dominate areas beyond the property boundaries it has been planted within. Thinning will keep the bamboo looking healthy and root barrier will present an obstacle to encroachment, but neither will prevent bamboo spread in the longer term. The key to successful bamboo maintenance is dealing with the rhizomes. Pruning rhizomes and ongoing vigilance for encroachment of rhizomes beyond containers and over root barriers is crucial. With bamboo being so prone to invasive behaviour, it must be kept in check. Bamboo should never be planted where the rhizomes have free access across a boundary line; nor should they be planted too close to boundaries.

12 Conclusion

Not all bamboos create problems, though those that do can cause significant damage and nuisance. Running bamboos are more likely to cause an issue, especially in more confined spaces (a small to medium-sized residential garden, for example). Nevertheless, even some clumping species can cause difficulties if not properly managed. When choosing what to plant, it is tempting to opt for bamboo because of its speed in achieving a tall screen. However, the taller a plant is, the greater the root or rhizome system needs to be in order to support it. Therefore, taller bamboos are more likely to produce a larger rhizome system that will spread across property boundaries, under patios or terraces, into lawns and, on occasion, into buildings.

To live in harmony with bamboo:

1. Choose the right species for the location. If the purpose is to create a screening hedge, choose any non-bamboo species without rhizomes. If bamboos are to be planted, choose less vigorous species and pay particular attention to points 2 to 5 below (see also 'Bamboos and gardening' and 'Useful websites' sections of Further Reading).
2. Install containment measures. Smaller bamboos should be containerised. Larger bamboos will require a properly installed root barrier. Be aware that many bamboos are capable of escaping from containment and so continual vigilance is crucial.
3. Ensure access is available around the entire plant. Do not locate the bamboo immediately adjacent to a wall or fence. Maintenance requires the ability to cut off rhizomes wherever they may spread. This is especially important in the region between the bamboo and a property's boundary.
4. Be scrupulous about maintenance. The less a bamboo is controlled, the more of a nuisance it is likely to create. Thinning, cutting and edging should be a vital part of a maintenance regime.
5. Be vigilant. Do not be complacent – not only regarding bamboo within property boundaries but also in regard to potential encroachment from outside sources. If tall bamboo is growing close

to a boundary on adjoining land it may only be a matter of time before it crosses the boundary. Do not allow bamboo to establish itself in this way.

When tackling bamboos, the importance of creating a Management Plan tailored to the specifics of the site should be emphasised. If an invasive bamboo is still immature, control and containment measures may be feasible. If the bamboo is well established and is causing a number of issues in regard to encroachment, amenity loss and a detrimental impact on the built environment, herbicide treatment or full removal may be the only options available.

The unconsidered planting of an invasive bamboo can cause damage, stress and disputes between neighbours. Whether it is the gradual encroachment into a lawn or shrub bed, or whether it is something more substantial, such as the dangerous lifting of paving slabs, the weakening of a boundary wall or the invasion into a dwelling itself, bamboos can have a significant effect on people's lives.

The authors believe that more consideration needs to be given to the suitability of problematic bamboo species for amenity landscaping. Adding invasive bamboo species to a planting scheme may lead to quite serious consequences for the home owner and the local environment. We therefore believe that more information about this tendency should be made available to a buyer at the time of purchase in clear and unambiguous language. Currently, information on these plants' invasive habits at the point of sale is, in our view, often lacking emphasis, unclear or even misleading. These can create unnecessary problems after a few years. Disputes with neighbours or environmental damage often are the result. The authors believe we need to find a way to remove the most troublesome species from our shores, so that we can all live in harmony with the less problematic bamboos that remain.

Appendix 1 – Herbicide application and equipment

Herbicide application must be conducted correctly in order for the herbicide to be effective. There are three main considerations to take into account when applying a herbicide to a target organism. These are:

- Dose;
- Timing;
- Method of application.

Failure to consider any of the above factors will result in a reduction of a programme's efficacy.

Regarding bamboo, the most suitable herbicide will almost certainly be a glyphosate-based herbicide, such as Roundup Proactive or similar. Early season applications in the spring are not advisable, as the growth and movement within the plant at this point is upwards and this will reduce the movement of the glyphosate into the bamboo's rhizomes.

One or two applications of herbicide (foliar applied) during the summer and autumn period are the best recommendation. The use of a long extending lance is advisable, especially on bamboo species where the growth exceeds two metres in height (Berthoud, as an example, make lances that can extend up to 3.6 metres in length). If applying herbicides at height, consideration must be given to spray drift. Spray drift increases with nozzle height, meaning drift onto non-target organisms becomes a very high risk if not managed correctly. Therefore, if a long lance is used (see Figs. 9.21, 11.1 & Ap1), a drift reducing nozzle should be employed. One suitable nozzle might be Billericay Farm Services' air-inducing hollow cone nozzle, which operates at low pressures (and so is suitable for a knapsack sprayer) and reduces drift satisfactorily. No application should take place in strong windy conditions, as this will cause unacceptable spray drift.

Figure Ap1 *Applying herbicide to bamboo using an extended lance and a knapsack sprayer.*

During the first winter period of the treatment programme, it is advisable to cut the culms down to approximately 10 to 15 cm long (or to just below the second node from the ground) using a sturdy pair of loppers. Following this operation, a solution of a glyphosate-based herbicide can be injected into the *in situ* stumps of the hollow culms from above (for maximum effect this should be carried out within four hours of cutting). For this operation, a stem injection gun can be utilised without using a needle: simply insert the gun tip over the culm stump and inject the required solution. For larger bamboo areas, the use of a blue spray dye is recommended to ensure culms are not missed or overdosed. A suitable injector gun would be the JK1000 stem injection tool supplied by Stem Injection Systems (see Figure Ap2), which is easily calibrated and can be used without a needle.

Figure Ap2 Using a stem injection gun without a needle: a technique pioneered by Brian Taylor in early 2014 for treating bamboo.

In subsequent years, follow-up visits will be required to monitor and treat any regrowth. This would typically involve foliar applications of herbicide to any regrowth during the summer or early autumn months. It may be necessary to expose live rhizomes and weed wipe these with a glyphosate-based herbicide. A sponge is a readily-available means of applying herbicide to exposed rhizomes.

All dose rates should be obtained from the product label. Always use herbicides in accordance with the label recommendations. Any queries or questions about herbicide use should be directed towards a BASIS-qualified person.

Appendix 2 – Tips for identifying the bamboo on site

It is not the purpose of this book to instruct on the identification of bamboo species, as this is a complex area. However, the authors can highlight what information needs to be recorded in the field to enable an identification to be made, as well as provide some key pointers to help distinguish between different bamboo genera.

As a first step, check whether the infestation is a single bamboo species or whether it is, in fact, several species. It is not uncommon to find bamboo hedges consisting of two to three different bamboo species. For reliable identification it is important to select mature culms, as young stems can often be unidentifiable.

For each species present, record the following:

o   Is the bamboo species running or clumping?
o   What is the maximum height of the culms?
o   What is the colour of the bamboo culms? (take photographs).
o   Does the bamboo branch at the nodes? How many branches at each node (average)?
o   Take representative samples of the culms and leaves for later identification.
o   Measure representative samples of the leaves' length and breadth – take these from well above mid-height of the culm.
o   What is the mature culm diameter?
o   What is the distance between nodes on the mature culms?
o   Are there any visible flowers? Take photographs/samples if so.
o   Are the culms smooth or ridged?
o   Are the culms furry?
o   Record culm sheath details – take photographs or samples.
o   Where the rhizomes can be exposed, are roots attached to the rhizome nodes only or are they attached across the rhizome at both nodes and internodes?

Genus	Leaves	Culm sheaths	Culms and branching
Arundinaria	Up to 20 cm long. Leaf blades ribbed. Culm sheaths very persistent.	Tardily deciduous or persistent.	Main culms up to 4 m, rounded in cross-section; nodes mostly with 3-7 branches.
Chimonobambusa	1-3 cm wide, minutely hairy when young. Mostly with 8-14 veins on either side of midrib.	Usually papery, tardily deciduous or persistent.	Main culms more or less square in cross-section, up to 8 m high; nodes mostly with 3 branches.
Fargesia	5-15 mm wide with 3-4 veins on either side of midrib; glabrous.	Deciduous.	Forms dense clumps with main culms more than 3 m high, nodes with 3 or more branches, typically 4-5.
Himalayacalamus	Delicate, without cross-veins.	Papery, deciduous.	Main culms 7 or 9 m. Nodes of mid-region of main stems from one to many branches, usually 15 in first year. Culm internodes increase progressively in length from the base.
Phyllostachys	About 10 cm long, variable. Sheds culm sheath quickly.	Remain attached for only a short time.	Culm grooved on alternate sides between branched nodes. Main culms 3-5 m high, nodes of mid-region mostly with two unequal branches and often a very small third one.
Pleioblastus	20 cm long, 3-30 mm wide; 2-7 veins either side of midrib; usually glabrous, rarely hairy.	Persistent.	Main culms up to 5 m or more, rounded in cross-section except for just above nodes, with one to many branches.
Pseudosasa	15-20 cm long, 2-4 cm wide, 5-9 veins either side of midrib. Culm sheaths persistent with leathery texture.	Leathery and very persistent.	Main culms up to 5 m high, round in cross-section except for just above nodes. Nodes mostly with one branch.

Genus	Leaves	Culm sheaths	Culms and branching
Sasa	4-5 times as long as wide, typically 2-5 cm wide. 6-14 veins either side of midrib; glabrous. Tessellated. Culm sheaths persistent.	Very persistent.	Main culms 0.5 m up to 3 m high, rounded in cross-section except for just above nodes. Nodes mostly with one branch.
Sasaella	1-3 cm wide, 8-20 cm long; 3-5 veins either side of midrib. Underside of leaves with short, soft, downy hairs; upper side sparsely hairy.	Persistent.	Main culms 0.5 m up to 1.5 m, rounded in cross-section except for just above nodes. Nodes mostly with one branch.
Semiarundinaria	1.5-4 cm wide, glabrous. Mostly with 4-6 veins on either side of midrib.	Remains attached for only a short time.	Forms dense clumps with main culms more than 3 m. Culms often flattened or grooved but only for upper internodes. Nodes of mid-region of main stems mostly with 3-5 branches.
Shibataea	10 cm long; 2.5 cm wide.	Papery.	Small in stature with main culms up to 2 m high. Culm internodes conspicuously flattened on one side. Nodes of mid-region of main stems swollen and very prominent; mostly with 4-5 branches. Culms tend to zig-zag.
Thamnocalamus	15 cm long; tessellated.	Not persistent.	Main culms up to 8 m or higher. Nodes of mid-region of main stems mostly with 2-3 branches increasing to 7. Culms often waxy; new culms often with blue-grey bloom.
Yushania	5-12 mm wide, mostly with 2-4 veins on either side of midrib. More or less glabrous.	Persistent.	Main culms more than 3 m and up to 13 m high, rounded in cross-section except for just above nodes. Widely spaced nodes.

Table A2a *Key features of bamboo genera commonly found in Great Britain and Ireland.*

The bullet points on page 85 are only a guide. Anyone with an interest in the subject, or is likely to need to survey bamboo, may wish to develop their own list of recording characteristics to enable an accurate identification.

Once a range of information has been gathered during your survey, either use a suitable IT-based resource or a published identification guide to assist in the identification of your bamboo species. There are two types of IT-based resources. First, there are websites that provide a comprehensive list of characteristic features with plenty of drawings or images of different types of culms, nodes, internodes, sheaths, branches, leaves and rhizomes. Examples of such websites can be found in the Further Reading section of this book. Second, there are apps such as Flora Incognita, Google Lens, Plant.id and PlantNet, which can be used to identify plants automatically by uploading your images to them. The apps listed are all free to use. Examples of a number of published identification guides can be found in the Further Reading section.

It is advisable to spend some time using different identification resources to determine which work best for you. You may find no single source will 'fit all', and that you are best combining an app that contains a published identification guide with the images provided on a comprehensive website. For assistance, Table A2a above provides the key features of the main genera of bamboos. Please refer to the Glossary for an explanation of technical terms.

Further Reading

Bamboos – general reading and information

Farrelly, D. 1984. *The Book of Bamboo*. Thames and Hudson, London.

Lucas, S. 2013. *Bamboo*. Reaktion Books, London.

Meredith, T.J. 2009. *Pocket Guide to Bamboos*. Timber Press, Portland, Oregon.

Ohrnberger, D. 1999. *Bamboos of the World*. Elsevier, Amsterdam.

Recht, C. & Wetterwald, M.F. 1992. *Bamboos*. B.T. Batsford, London.

Vorontsova, M.S., Clark, L.G., Dransfield, J., Govaerts, R. & Baker, W.J. 2017. *World Checklist of Bamboos and Rattans*. Science Press, Beijing.

Identifying bamboos found in Great Britain and Ireland

Chao, C.S. 1989. *A Guide to Bamboos Grown in Britain*. Royal Botanic Gardens, Kew.

McClintock, D. 1980. 'Descriptive key to bamboos naturalised in the British Isles.' *Watsonia*, 13, 59-61.

McClintock, D. 1992. 'The shifting sands of Bamboo genera.' *Plantsman*, 14, 169-177.

Poland, J. & Clement, E. 2009. *The Vegetative Key to the British Flora*. John Poland, Southampton in association with the Botanical Society of Britain and Ireland, Durham.

Rickel, C. 2013. *Field Guide to Identification of Phyllostachys Invasive Running Bamboo*. Institute of Invasive Bamboo Research, USA (on line).

Ryves, T. B., Clement, E. J., Foster, M. C. 1996. *Alien Grasses of the British Isles*. Botanical Society of the British Isles, Durham.

Sell, P.D. & Murrell, G. 1996. *Flora of Great Britain and Ireland*. Volume 5. *Butomaceae – Orchidaceae*. Cambridge University Press.

Stace, C. 2019. *New Flora of the British Isles*. Fourth Edition. Cambridge University Press, Cambridge.

Bamboo identification website (www.bamboo-identification.co.uk)

Lewis Bamboo website (https://lewisbamboo.com)

Guadua Bamboo website (www.guaduabamboo.com)

Biology and ecology of bamboos

Gucker, C. L. 2009. *Phyllostachys aurea*. Fire Effects Information System, US Department of Agriculture.

Stapleton, C. 1994. 'Form and function in the bamboo rhizome.' *Journal of the American Bamboo Society*, 12, 21-29.

Bamboo control

Anon. 2011. *Notes on the Control of Bamboo*. Monsanto UK Ltd, Cambridge.

Fennell, M., Jones, L. and Wade, P.M. (compiled by) 2018. *Practical Management of Invasive Non-Native Weeds in Britain and Ireland*. Packard, Chichester.

Royal Horticultural Society. *Bamboo control*. (https://www.rhs.org.uk/advice/profile?pid=210)

Bamboos and gardening

Anon. 2002. *Grasses and Bamboos*. RHS Practicals. Gardners Books, Eastbourne.

Bell, M. 2000. *The Gardener's Guide to Growing Temperate Bamboos*. David & Charles, Newton Abbot, Devon.

Costello, L. 2008. *Success with Bamboos and Ornamental Grasses*. Guild of Master Craftsmen Publications, Lewes.

Lawson, A.H. 1968. *Bamboos. A Gardener's Guide to Their Cultivation in Temperate Climates*. Faber and Faber, London.

Meredith, T.J. 2001. *Bamboo for Gardens*. Timber Press, Portland, Oregon.

Useful websites

Bamboo Inspiration website (www.bamboo-inspiration.com)

Bambu Batu: The House of Bamboo website (www.bambubatu.com)

Bamboo Botanicals (www.bamboobotanicals.ca)

The Bamboo Garden website (http://bamboogarden.com)

Gardening know how (www.gardeningknowhow.com)

Jacksons Nurseries website (www.jacksonsnurseries.co.uk)

Klodd, Annie, and Curran, Bill (2016). *Integrated Weed Management*. Penn State University. (https://growiwm.org/wp-content/uploads/2016/10/Integrated-Weed-Management-fact-sheet.pdf)

Royal Horticultural Society website (www.rhs.org.uk)

USDA National Conservation Resources Service website (www.plants.usda.gov)

List of illustrations

List of tables

Glossary

Aerial root or rhizome: Roots/rhizomes that grow above ground level.

Amphimorph: Bamboos that demonstrate both clumping and running characteristics at the same time. Very rare.

Annual flowering: A plant that produces flowers every year.

Auricle: Ear-like flaps located on the upper part of the sheath on both sides of the blade.

Bamboo: A number of evergreen, perennial, flowering species that are included within the sub-family Bambusoideae, of the grass family Poaceae.

BASIS: Originally the acronym , but now the trading name of the organisation that trains practitioners and maintains high standards for use and storage of agro- and environmental pesticides and herbicides.

Blade: The broad section of a leaf that is dedicated to photosynthesis. It is typically flat in bamboos.

Bloom: Fine waxy-looking powder, usually white, covering part or all of the plant.

Canes: Dead bamboo culms.

Casual: A non-native plant species growing unaided by direct human intervention (i.e. not planted) but not yet permanently established.

Clumping bamboo (see also Pachymorph): Bamboos with short rhizomes that turn up and form new culms almost immediately. They tend to be tightly clumped and are generally considered to be less invasive in nature.

Columnar: A cylindrical object or pillar that does not vary in thickness for most of its length (e.g. a bamboo culm).

Culm: A jointed and hollow grass stem (e.g. a bamboo stem). Culms in bamboo are typically large and woody; less so in other grasses. Leaves and side branches may emerge from the nodes. The circumference of new culms is thicker each successive year than the previous years' culms. This process continues until an optimum thickness for the growing conditions is achieved.

Culm leaf: A leaf that may be found at the end of an emerging bamboo shoot. It offers some protection and support to the culm as it grows.

Cultivar – see Taxonomy

Dormancy: A period of inactive growth. Dormancy can be induced or be naturally occurring.

Efficacy: The effectiveness of a procedure or operation to control or limit growth.

Family – see Taxonomy

Foliar application: Diluted herbicide applied to the leaves of a plant, typically by means of a spray.

Genera/Genus – see Taxonomy

Glabrous: A smooth surface (e.g. of a leaf) devoid of hairs or down.

Glaucous: A blue-grey, white or silvery coating found on leaves and stems. More common on plants native to coastal regions.

Glyphosate: The active ingredient of many translocative non-selective herbicides (e.g. Roundup).

Gregarious flowering: A term used to describe a species of bamboo in which all clumps flower at the same time regardless of location.

Hardiness: A plant's ability to withstand cold temperatures. Plants termed to be 'tender' will not tolerate temperatures lower than 5°C. 'Half-hardy' plants tolerate temperatures down to 0°C. 'Frost hardy' plants tolerate temperatures down to -5°C. 'Full hardy' plants can tolerate temperatures as low as -15°C.

Herbicide: A class of pesticide used to eradicate or control plants.

Inflorescence: A term used to describe the arrangement of flowers on a plant.

Internode: The section of a culm between the nodes, which is hollow in almost all bamboos.

Japanese knotweed: (*Reynoutria japonica*) A known invasive non-native plant, listed in Schedule 9 of The Wildlife & Countryside Act.

Leptomorph: The name given to a bamboo with a spreading habit. Leptomorphs produce running rhizomes that typically grow horizontally and can spread great distances.

Monocotyledon: Flowering plants whose seedlings have only one seed-leaf (cotyledon). Leaves are usually narrow, unstalked, often parallel-sided and nearly always parallel-veined. The main families (see Taxonomy) of monocotyledon are the grasses, including bamboos, sedges, rushes and orchids.

Neck: To be found in clumping bamboos (pachymorphs), it is the thinner section linking older and younger rhizomes.

Node: Joints on culms and rhizomes that seal off sections of the stem/rhizome internally and from which new growth (i.e. leaves, stems, branching rhizomes) can emerge. Nodes are often swollen and on culms can be a different colour to the rest of the stem.

Pachymorph: A bamboo with a clump-forming habit. Rhizomes are short and produce aerial growth (culms) at their tips. The rhizomes tend to be thicker than the culm emerging from it, with new rhizomes developing from buds on existing rhizomes.

Perennial: A plant that takes three or more years to complete its life cycle. Most perennials will live for many more than three years before they decline from natural causes or die.

Propagule: Part of a plant that, if separated from its parent plant, can grow into a new one. Most typically rhizome fragments in bamboos, but can also include seeds.

Property Care Association (PCA): A Trade Association offering expert training in invasive weeds. (www.property-care.org)

Rhizome: A stem that usually grows horizontally underground and acts as a storage organ for starches and proteins. Rhizomes typically have true roots and new stems emerging at the node points.

Root: Part of the plant that usually grows downwards from the rhizome nodes. It absorbs water and nutrients from the soil and anchors the plant to the ground. In bamboos, roots are typically quite fibrous in appearance and are not segmented like culms and rhizomes.

Running bamboo (see also Leptomorph): An invasive type of bamboo that produces rhizomes that can spread long distances. New culms emerge from anywhere along the length of the rhizomes.

Schedule 9: Part of the Wildlife and Countryside Act (1981) wherein species of concern are listed.

Sheath: A leaf base that firmly encircles the stem and covers and protects the emerging new growth. On bamboos, the sheaths are often very noticeable, especially as they turn pale, papery and fall off when they dry out.

Sheath scar: A circular scar left beneath the nodes after the sheath has fallen from where it was originally attached to the culm.

Species – see Taxonomy

Sporadic flowering: A type of flowering that does not seem to be determined by environmental factors and does not occur at the same time as other members of the same bamboo population.

Stem injection: A method of introducing precise amounts of herbicide into hollow culms and stems. With bamboo, the culms are typically cut between the first and second nodes and the injection is made into the internode space from above.

Sulcus: A vertical groove running the length of the internode. In some species, the sulcus can be found on alternative internodes.

Taxonomy: An internationally shared means of biological classification, whereby organisms are placed into more and more inclusive groupings. The organisation from larger to smaller, more specific, categories is called a hierarchical system. All species are categorised by Domain, Kingdom, Phylum (Division), Class, Order, Family, Genus and Species. Bamboo belongs to the Family of Poaceae. The Family is divided into sub-families, with bamboo being in the sub-family of Bambusoideae. The Genus (plural: Genera) is similarly sub-divided, as there are approximately 120 different Genera of bamboo. The Species is the basic unit of biological classification. Individual bamboos are normally identified by both Genus and Species. There is a further distinction: Cultivar. A Cultivar is a selected or bred plant that differs from a typical member of the species. To take a common type of bamboo as an example:
Family: Poaceae
Sub-Family: Bambusoideae
Genus: *Phyllostachys*
Species: *Phyllostachys aureosulcata*
Cultivar: 'Spectabilis'

Tessellated: Divided into squares.

Weed wipe: A method of applying herbicide to leaves, stems and rhizomes, using a special weed-wiping device or a sponge soaked in herbicide (diluted in accordance with the label recommendations). A slow but more precise way of applying herbicide than spraying.

Wildlife and Countryside Act (as amended) 1981 (WCA): UK legislation which, among other things, identifies species of environmental concern which must not be caused to spread into the wild and, for some species, imposes a ban on sales. Bamboos are not (at the time of writing) included in this legislation.

Wildlife and Natural Environment (Scotland) Act 2011 (the WANE Act): Scottish legislation concerning the way land and the environment is managed in Scotland.

Index

About the Authors

Brian Taylor has an MSc in Crop Protection and an HND in Amenity Horticulture. He is a Certificated Surveyor in Japanese knotweed (CSJK), and was the Property Care Association (PCA) Student of the Year 2016; also a Member of the BASIS Professional Register (MBPR). While a teenager, Brian worked on Forestry Commission trials on invasive weeds, such as rhododendron, and later was an Assistant Agronomist on field trials for Monsanto. He spent six years as a self-employed researcher carrying out commercial evaluation of spray nozzles and machinery for drift categories. He began working  with invasive weeds, such as Japanese knotweed and giant hogweed, in 2004, which led to forming The Knotweed Company in 2010 where he has also developed management techniques for dealing with invasive bamboo since 2015.

Jim Glaister BA(Hons) CSJK and was the PCA Student of the Year 2015. He began working in the invasive weed industry in 2004, initially for Wreford Ltd and then freelancing for several invasive-weed companies before joining The Knotweed Company in 2013. He is a contributing author for the Royal Institution of Chartered Surveyors' (RICS) online subscriber information portal *isurve* since 2009. He frequently gives CPD presentations and webinars on a variety of invasive-weed topics to environmental consultants, local authorities, property management companies  and organisations such as the RICS and the PCA. He has been an examiner and marker for the PCA's CSJK examinations since 2015, as well as being a member of the PCA Education Committee and its Annual Conference Committee. With Brian Taylor he has developed a number of strategies for the management of Japanese knotweed and invasive bamboo.

Max Wade BSc(Hons) PhD CEcol CEnv FCIEEM is currently Technical Director (Ecology) at AECOM. He began his career in higher education at the University of Wales Institute of Science and Technology, later moving to Loughborough University where he helped set up the International Centre of Landscape Ecology (ICOLE), which conducted research and consultation first on invasive aquatic plants and later on land-based problematic invasive weeds such as giant hogweed, Himalayan balsam and Japanese knotweed. He was appointed Professor of Ecology and Head of  the Department of Environmental Sciences at the University of Hertfordshire, Hatfield. He has authored a number of manuals, field guides and scientific papers, and advises nationally on invasive non-native plants and animals. Max was Founding Chair of the Invasive Weed Control Group (120 member companies) of the PCA, 2012-18, and is President of the Chartered Institute of Ecology and Environmental Management.

Some other practical books about invasive weeds …

ISBN 9781853411656 £35.00   ISBN 9781853411274 £27.00